Ernest Rutherford

Maria Gill

Illustrated by Alistair Hughes

upstart press

Curiosity exists in all of us.

An ordinary country boy from Nelson, New Zealand would one day make major discoveries about how old Earth was, and answer centuries-old questions such as, 'What are atoms made of?' and 'How do they work?' Ernest Rutherford's three major discoveries and many minor ones helped shape much of modern science. His work began an explosion of discoveries around matter and physics.

Memorial statue of young Ernest Rutherford where he was born in Spring Grove (now called Brightwater).

At the end of Earth, in the tiny town
of Spring Grove, New Zealand,
a baby called Ernest Rutherford
gurgles in his bassinet.
It is 30 August 1871.

Ernest grows up to be an ordinary boy in an ordinary family of fourteen. He milks the cows, feeds the pigs, and collects the chooks' eggs.

Before he starts school,
his mother teaches him
to read and write,
and to learn his 12 times tables.

When the other children play
football at playtime,
Ern sits under a tree,
reading and daydreaming.

One day, Ern's teacher gives him a science book.
It inspires Ern to fire a cannon,
count how far away lightning strikes,
and pull apart a clock
to see how it works.

The experiments raise questions in Ern's curious mind.

To find answers, Ern dreams of going to high school.
But his family is too poor.
He must sit a scholarship exam.
But he fails.

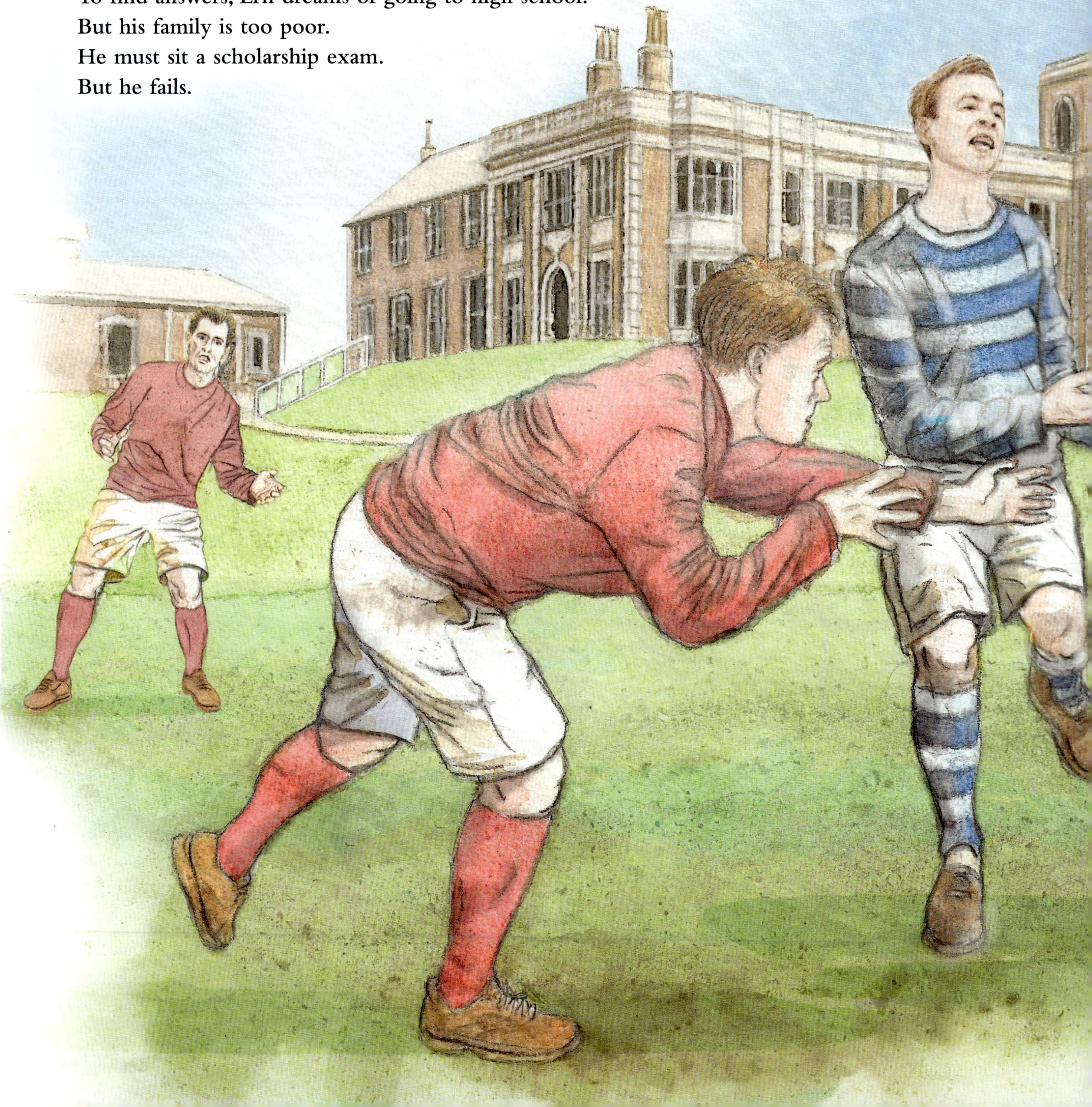

Ern doesn't give up.
On his second try, he succeeds.
Ern studies hard and plays
rugby for exercise.

After high school, he dreams of going to university
and sits another scholarship exam.
But he fails again.
Ern doesn't give up.
On his second try, he succeeds.
Lack of space at the university
forces Ern to experiment in
corridors, halls and basements.

After university, he dreams of
doing research overseas.
He submits his two science
projects, but Ern fails yet again.
He thinks all is lost.

Months later, while digging a potato patch
for his mother at home,
a telegram arrives.
The winner has withdrawn!
Ern has his scholarship.

He researches in English and
Canadian laboratories,
designing experiments to unlock
science's deepest mysteries.

Ern breaks the world record for sending a
wireless message over a long distance.
He also finds answers to centuries-old questions,
such as 'How old is Earth?'
Ern discovers that atoms are mostly empty space.
At their centre is a dense nucleus,
and atoms *can* be split or combined.

Ern travels the world giving talks, sharing the wonders of science.

He never forgets his homeland
and humble beginnings,
and marries his sweetheart
in New Zealand.

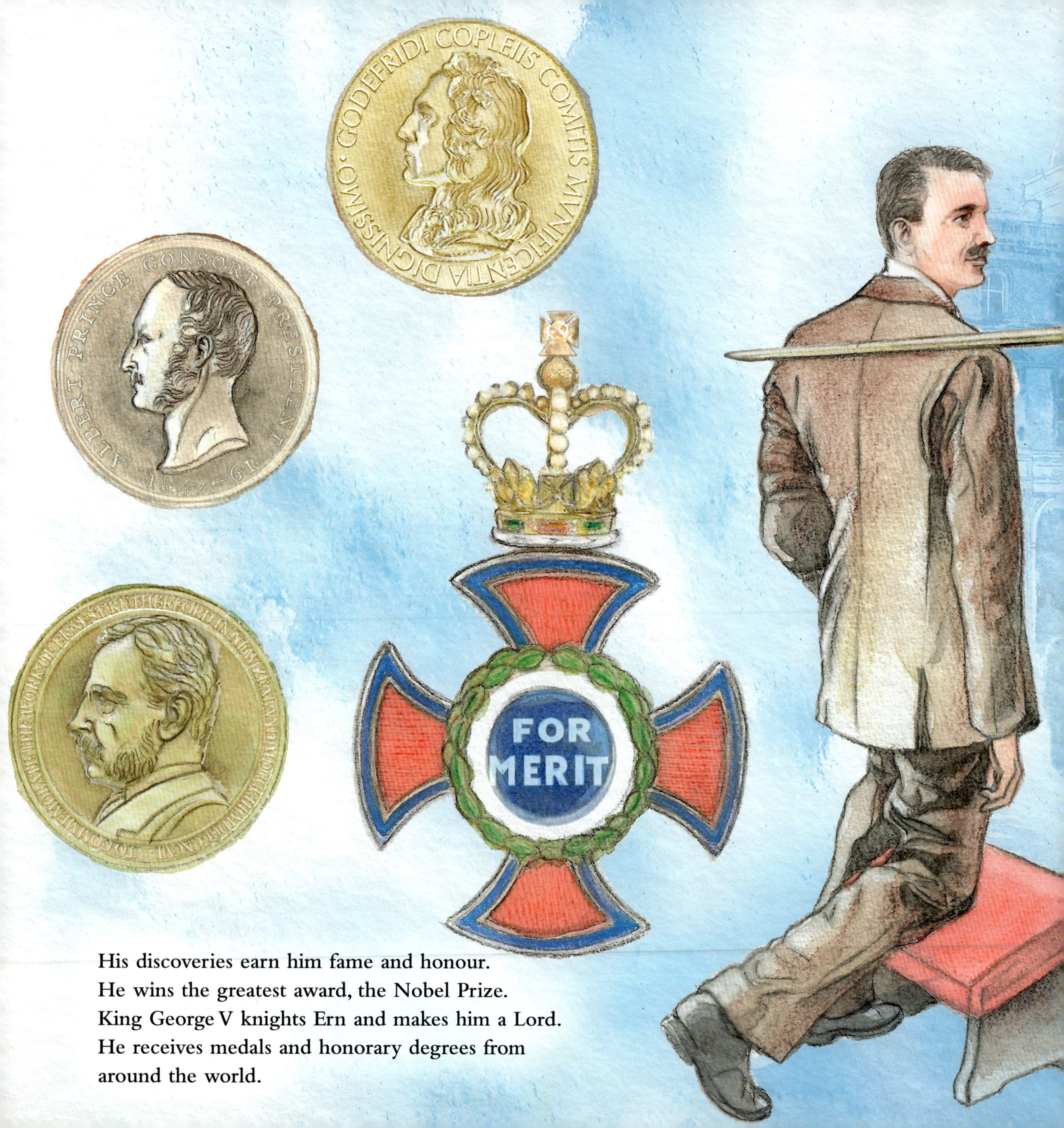

His discoveries earn him fame and honour.
He wins the greatest award, the Nobel Prize.
King George V knights Ern and makes him a Lord.
He receives medals and honorary degrees from
around the world.

NAT• MDCCC XXXIII OB• MDCCC XCVI
ALFR• NOBEL
THE•INSTITUTION•OF•ELECTRICAL•ENGINEERS
AWARD•OF•FARADAY•MEDAL•TO•ERNEST•RVTHERFORD
INVESTAS•VITAM•IUVAT EXCOLUISSE•PER•ARTES
NATURA
SCIENTIA
REG•ACAD• SCIENT•SUEC
E•RUTHERFORD MCMVII
ELLIOT CRESSON MEDAL
FOUNDED 1848

Ern is remembered for being one of the

greatest scientists.

Ernest Rutherford
is to the ATOM

and Albert Einstein
is to RELATIVITY

what Charles Darwin
is to EVOLUTION

Isaac Newton is
to MECHANICS

Marie Curie is to
RADIATION

That ordinary boy from a tiny town
at the end of Earth
proved that perseverance and curiosity
can give you the greatest gift of all . . .

KNOWLEDGE

Ernest Rutherford's memory lives on

Teachers considered Ernest Rutherford just an ordinary boy who studied hard to get top marks. But Ernest achieved his dream of furthering his education despite the obstacles. In those days, large families like the Rutherfords could not afford to send their children to high school and university. Students had to win scholarships to pay for their education. Each time Ernest failed, he tried again until he received his scholarship.

It was Ern's curiosity about the world, his 'can-do' attitude even with little money, and not giving up that made him into a highly successful scientist. Ern was not interested in making money from his discoveries. Instead, he wished to add to humankind's knowledge of how things worked.

When Ern died at sixty-six years old, he'd accomplished much more than he had ever hoped or imagined. His ashes lay in Westminster Abbey alongside the world's other great scientists, including Sir Isaac Newton's memorial, Michael Faraday, and Charles Darwin.

He achieved immortality with his breakthroughs in science and is one of our greatest-ever New Zealanders.

His name endures:

13 stamps from 10 countries

Statue of a golden-shaped man splitting apart the atom at Qingpu Statue Garden, Shanghai, China

An asteroid, a crater on Mars and the Moon, a mineral, and a mountain

Ernest Rutherford's experiments

Over a century ago, Ernest Rutherford made his greatest discoveries while working in laboratories in England and Canada.

❶ In New Zealand he invented a detector of very short pulses in circuits and when he went to England, he adapted it for wireless signals.

❷ Ern showed that radiation comprised two types of rays, which he called alpha and beta. Radiation helps benefit humankind in medicine, agriculture, space exploration, mining, and generating electricity.

❸ Ern discovered that some heavy elements spontaneously disintegrate on their own. It helped science date the age of the Earth, trees, and ancient objects.

❹ His experiments also showed that the atom can be forcibly split.

❺ He identified the nuclear structure of the atom. This is the research for which he is most well-known. Understanding the make-up of atoms has led to significant advances in physics and chemistry.

Ernest Rutherford's timeline

1871
30 August: born to parents Martha and James Rutherford in Spring Grove

1892
Graduates with a Bachelor of Arts

1893
Graduates with a Master of Arts, double first-class Honours in Mathematics and Physics

1895
Gains a Bachelor of Science degree and wins a research scholarship

1901
Graduates Doctor of Science from the University of New Zealand. Daughter Eileen Rutherford born

1904
Awarded the Rumford Medal by Britain's Royal Society

1908
Invents the Rutherford-Geiger tube. Awarded Nobel Prize in Chemistry

1914
King George V knights Ernest Rutherford

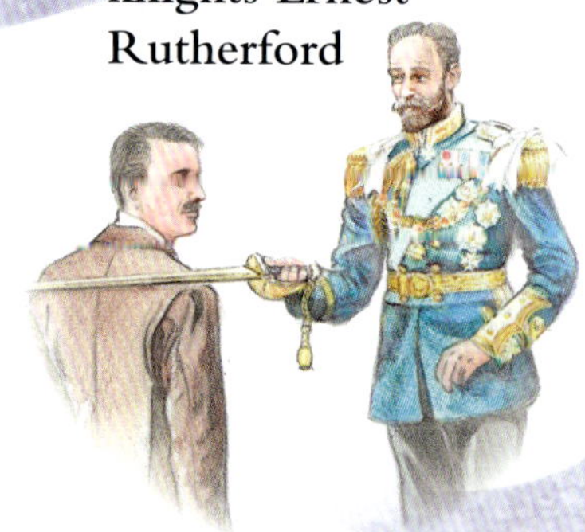

1896
Transmits wireless signals over 800 metres using his magnetic detector

1898
First to coin the terms alpha and beta rays

1899
Demonstrates the principles that are the basis of modern smoke detectors

1900
Marries Mary Newton in Christchurch

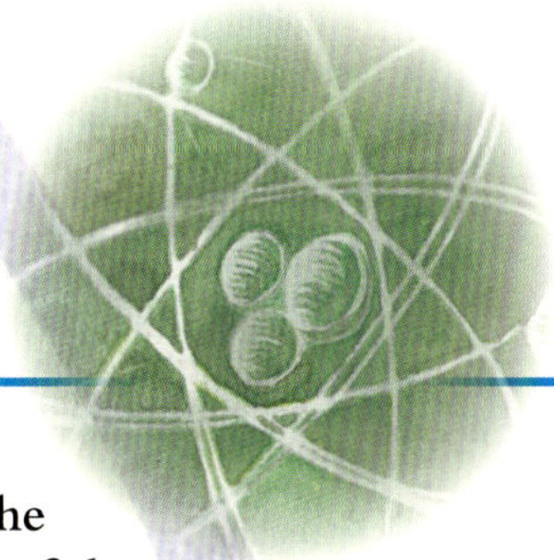

1915–1917
First World War: develops methods for the underwater detection of submarines

1920
Predicts the existence of the neutron in the atom

1931
Made Baron Rutherford of Nelson

1937
Ernest Lord Rutherford dies on 19 October, aged 66 years

GLOSSARY

Alpha particle: (α) Also called alpha rays or alpha radiation, it comes from the decay of the heaviest radioactive elements such as uranium and radium. Positively charged alpha particles comprise two protons and two neutrons.

Atom: The smallest unit of ordinary matter that forms a chemical element such as gold, uranium, or carbon. Each atom has protons and neutrons in its centre (the nucleus), and electrons.

Beta particle: (β) Also called beta ray or beta radiation, it is a small, fast-moving particle with a negative electrical charge that emits from an atom's nucleus.

Nucleus: Is the positively charged central core of an atom, comprising protons and neutrons.

Protons: These are the positively charged particles found within an atomic nucleus.

Radiation: Is energy that comes from a source, such as ultraviolet (UV) from the sun, x-rays, or alpha particles emitted from radioactive decay of uranium.

A catalogue record for this book is available from the National Library of New Zealand

ISBN 978-1-990003-68-4
An Upstart Press Book
Published in 2023 by Upstart Press Ltd
26 Greenpark Road, Penrose, Auckland 1061
New Zealand

Design and Illustration by Alistair Hughes

Printed by Everbest Printing Co. Ltd., China

Bibliography

Rutherford: Scientist Supreme (1999)
by John Campbell. AAS Publications, Christchurch.

Rutherford: The early years (1991)
by Ian Cox and Mike Whittall.
Nelson Teaching Resource Centre.

www.rutherford.org.nz (2001)
by John Campbell

Rutherford's Den, The Arts Centre, Christchurch

Acknowledgements

Thanks to the Arts Centre in Christchurch for the Creative Arts Residency, which enabled Maria Gill to research the book at Rutherford's Den. Also, thanks to Dr John Campbell for his expert help with the manuscript and Jenny Stuart at Rutherford's Den for fact-checking the final manuscript.